Know M1 Garand Rifles

by E.J. Hoffschmidt

Blacksmith Corp.

CONTENTS

	Page
HISTORY	5
CARBINES AND SHORT RIFLES	8
TABLE OF "E" NUMBER GUNS	10
SNIPER MODELS	14
M1 PRODUCTION	18
FULL AUTOMATIC GARANDS	21
REMINGTON M1 CONVERSIONS	23
MAGNESIUM STOCK M1	26
TEFLON FINISHED M1	26
NATIONAL MATCH M1	27
TABLE OF "T" NUMBER GUNS	31
M14 HISTORY	42
NAVY CONVERSION	46
ORDNANCE UPGRADING PROCEDURE	47
M1 & M14 SERIAL NUMBERS	66
GARAND COPIES	67
FUNCTION OF M1	70
TROUBLESHOOTING CHART	73
TAKEDOWN	75
AMMUNITION	79

© 75 BLACKSMITH INC.
All Rights reserved under International and Pan American Copyright Conventions
ISBN 0-941-54002-2

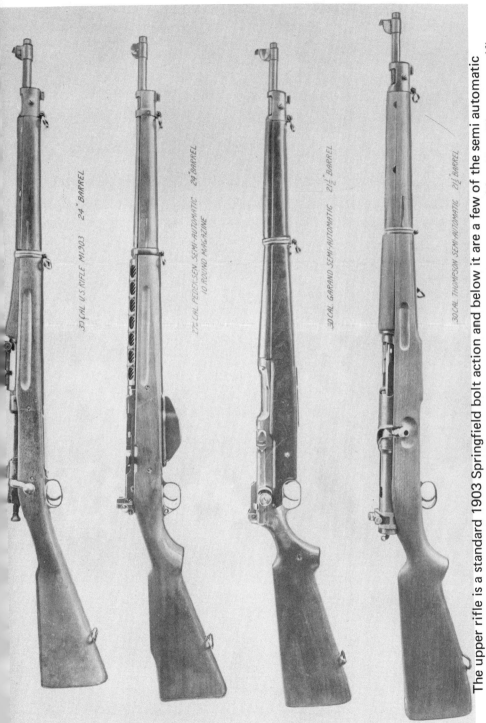

The upper rifle is a standard 1903 Springfield bolt action and below it are a few of the semi automatic rifles that sought to replace it, the 276 Pedersen, the 30 caliber Garand and the 30 caliber Thompson rifle.

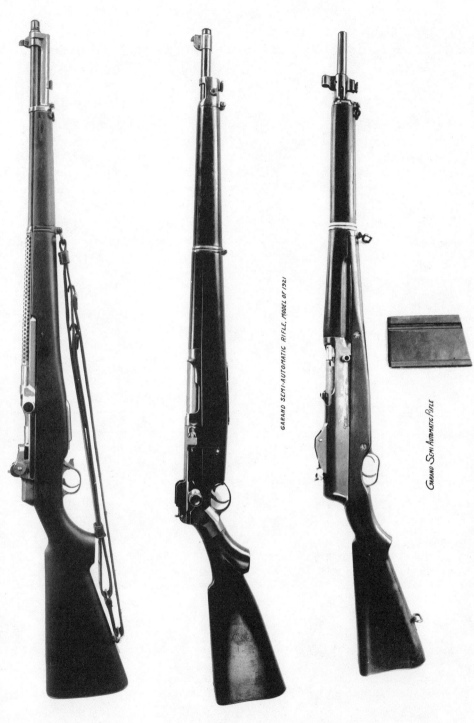

GARAND SEMI-AUTOMATIC RIFLE, MODEL OF 1921

GARAND SEMI-AUTOMATIC RIFLE

John Garand made a number of experimental rifles before the design was finalized. The upper gun in 276 caliber was tested in 1931 and recommended for adoption in 30 caliber by General Douglas McArthur.

HISTORY

There were dozens of other self loading rifles around long before the M1 Garand. They all worked with varying degrees of efficiency and utilized a wide variety of operating systems.

In 1921 after the results of WW I had been partially digested, the United States sponsored a series of competitive tests on existing rifles. The list of specifications called for a weapon of not less than .25 cal and not more than 30 cal preferably to fire the 30 cal M1906 cartridge. Furthermore, the Army requested that it be simple, strong, compact and reliable. The mechanism was to be well protected from sand, rain and dirt. It was to be magazine fed with a capacity not to exceed 10 rounds. The desired rifle was to be strictly semi automatic; not over 10 lbs and have a balance and outline similar to the M1903.

After the above characteristics were established a number of guns were submitted.

Prior to the 1921 tests, the Engineering Division of the Ordnance Dept. had tested a number of semi automatic rifles. Among the rifles tested was the Dommarito automatic rifle. This rifle was recoil operated and had a toggle breech something like an overgrown Luger pistol. The Swiss Rychiger was tested and found wanting because it would only function when well oiled. The Bang rifle came close to what the Army was looking for. It worked well but it required about one hour to strip.

Major Elder took the best features from the Bang rifle and the Rychiger, he threw in a Mannlicher bolt action and came up with another "unusual" design.

The guns of the 1921 test series brought forth a new batch of potential US service rifles. A revised Bang rifle was submitted. The U.S. Machine Gun Company submitted a Berthier rifle with a gas system that was later used on the M14 rifle. Colt and Thompson both submitted delayed blow back auto loading rifles. Three European rifles were also tested; a Belgian, a Czech, and a German gun.

John Garand submitted a very unique rifle that required a special cartridge. The cartridge primer pushed back a piston around the firing pin that unlocked the gun. J.D. Pederson, designer of the "Pederson Device" of WW I and the Remington automatic pistols submitted a very complicated rifle that used a locking system similar to the Luger pistol or Maxim machine guns.

Needless to say, none of these other ingenious designs filled the bill.

The M1 Garand was not a revolutionary weapon. It was the product of years of design and redesign. Years of experimenting with various systems of operation, testing new ideas and refining old ones.

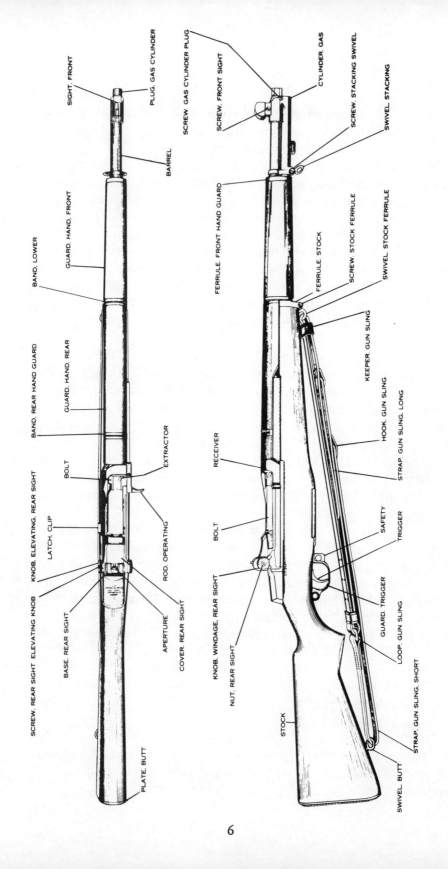

Finally after numerous battles with Congress, funds were appropriated for the new rifle. Luckily for our side, there were enough cool heads in Congress to give America one of our most important weapons of WW II. In January 1936, years before Hitler started in the Real Estate business, we adopted the most advanced military infantry rifle in the World. The Russians had adopted their semi auto rifles about the same time, but neither the Simonov or the Tokarev could hold a candle to the M1 Garand in rugged reliability.

Melvin Johnson designed a rifle that came the closest to eliminating the Garand. It was a recoil operated semi auto rifle with an unusual rotary magazine that could be loaded with loose cartridges without interfering with the round in the chamber. It too fell by the wayside although it was used by the Marines to a certain extent during the early days of WW II.

Early in 1941, the U.S. Marine Corps finally adopted the Garand.

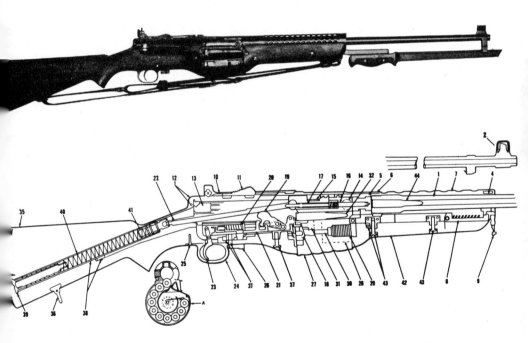

The photo above shows the Johnson type "R" semi automatic rifle fitted with a 20 inch sword bayonet. The blade extended 11 inches beyond the muzzle. Below, the cutaway clearly shows how the 10 round rotary magazine fed the cartridges to the bolt.

GARAND CARBINES OR SHORT RIFLES

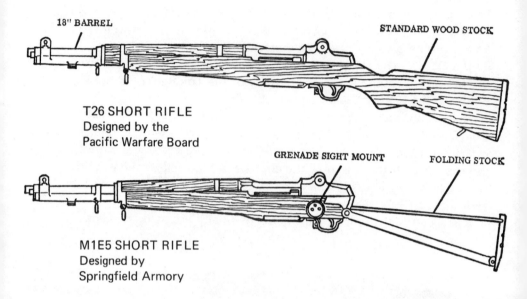

Carbines have always had an appeal to the average foot soldier. They are lighter and handier than the usual full size rifle.

Reports from the farflung combat theaters in WW II indicated the need for a carbine version of the M1 Garand. The 30 cal M1 carbine was just not powerful enough for some combat situations.

In 1944 the Ordnance Dept directed Springfield Armory to fabricate a test model with an 18 inch barrel and a folding stock similar to pantograph stock used on the M1A3 carbine. This new gun was designated the M1E5. Tests at Aberdeen indicated that accuracy up to 300 yards was comparable to the standard M1 but the muzzle blast and flash were excessive. Tests also indicated the need for a pistol grip on the stock.

In July of 1945 Col. William Alexander, President of the Pacific Warfare Board requested 15,000 short M1 rifles. To help speed things up, the Pacific War Board had the Ordnance unit of the 6th Army in the Philippines make up 150 short rifles. They sent one by courier to the Ordnance Dept. They in turn put the M1E5 on a shortened stock and called the gun the T26. 15,000 were scheduled for production, but VJ day ended the need.

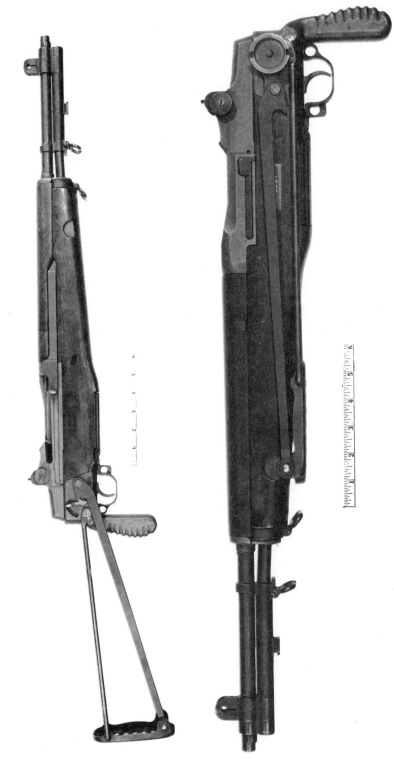

The photos above show the M1E5 carbine developed by Springfield Armory in 1944. The pistol grip was added after the initial test. Later the folding stock was dropped and a full wood stock was substituted. This gun was listed as the T26.

WARTIME M1 RIFLES

Combat conditions in all theaters gave the M1 Garand a quick harsh shakedown. Actually few major problems turned up. The first major problem was the unstable rear sight. After several hundred rounds, the rear sight developed a tendency to jump. The rear sight pinion was revised and the problem controlled until a new sight assembly was issued in October 1944.

The M1 Garand had another problem that plagued it thru its short history. Under certain conditions, usually extremely heavy rain, the action froze after firing only a limited number of rounds. Various lubricants were tried. Lubriplate 130-A seemed to work best, but numerous mechanical fixes were tried.

M1E1

The M1E1 was the first of the mechanical fixes that tried to correct the "frozen" action problem. The fix consisted of changing the cam angles on the rod to lessen the freezing or binding of the action. It did not work too well.

M1E2

This designation was assigned to the first sniper version and is covered more fully on Page 15.

M1E3

In a further attempt to correct the frozen action problem, a roller was added to the cam lug on the bolt. The operating rod was redesigned to accommodate this inovation. It worked fairly well, but the duration of the stoppages varied from gun to gun. This fix was not incorporated into the M1 but was introduced into the full automatic version, namely the T20E2.

M1E4

Further attempts at improving the "galling effect" that added to the action freeze-up resulted in the M1E4. This experimental M1 used a gas cut-off, gas expansion system. It resulted in a lower operating rod velocity and resulted in cutting the bolt and rod impact by half. It also kept the operating rod and spring in much longer contact with the hot barrel gasses. This resulted in excessive heating of the above components. Scratch another good idea.

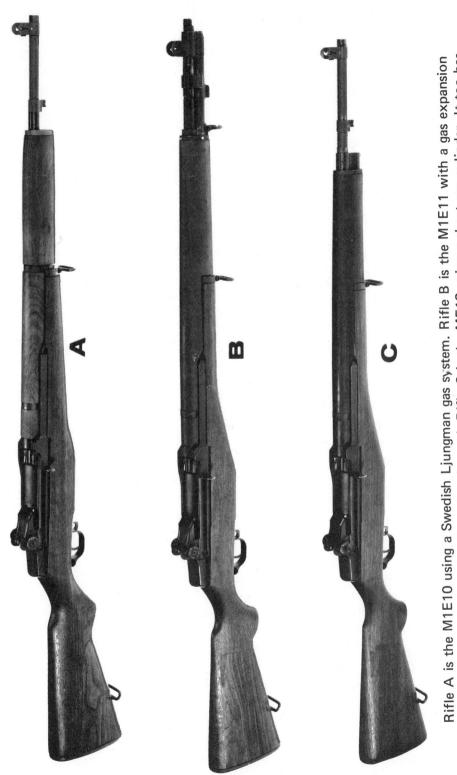

Rifle A is the M1E10 using a Swedish Ljungman gas system. Rifle B is the M1E11 with a gas expansion system and a one piece aluminum hand guard. Rifle C is the ME12 using a short gas cylinder. It too has a one piece aluminum hand guard. They all fire the 30 caliber M1 cartridge.

M1E5

A demand for a short rifle or Carbine Garand prompted the Army to build the M1E5. (See pages 8 and 9.)

M1E6

When the M1E2 prismatic telescope sight did not pan out, Springfield tried a side mounted scope with an A3 Springfield rear sight. This model required too much modification of the M1 receiver so the idea was dropped.

M1E7

The M1E7 was the prototype of the M1C Sniper Garand. A sturdy Griffin and Howe telescope mount was screwed and pinned to the receiver. A commercial Lyman Alaskan telescope was fitted and given the designation M73.

M1E8

The M1E8 was the prototype of the M1D Sniper Garand. It became the substitute standard sniper rifle in 1944. (See Pg 15–16.)

M1E9

In order to eliminate the heat problem of the M1E4, a long tappet principle was developed. The piston moved 1½ inches to the rear imparting a blow to the operating rod. Inertia did the rest — carrying the operating rod to the rear. The production status of the M1 did not warrant the design change so the idea was held for later use.

M1E10

The Swedish Ljungman rifle supplied the gas system for the M1E10. A long tube extended from the gas port to a support 2/3 of the way down the barrel and terminated into a piston shape. This short piston entered about 1-inch into a projection on the operating rod. When the gun fired, the gasses pushed directly on the operating rod. This idea was dropped because of heat problems.

M1E11

This gun was basically an M1 with a gas cutoff and expansion system built into the standard M1 gas cylinder casting. The gas port was moved back about 3 inches. A one piece aluminum hand guard coated with a heat resistant epoxy replaced the wooden two piece hand guard.

M1E12

The M1E12 went back to a short gas cylinder with the gas port about 6-inches from the muzzle. It had a one piece coated aluminum hard guard similar to the E11 gun and fired the standard 30.06 cartridge.

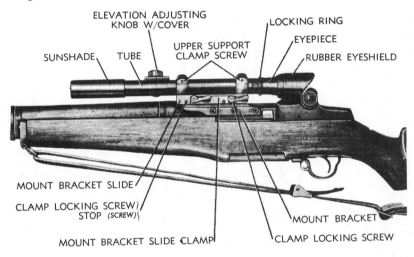

The M1E7 was the prototype of the M1C Sniper Rifle.

In the early 1950's, the U.S. Marine Corps ordered a quantity of Stith 4X type telescopes from the Kollmorgen Optical Corp. for use on their M1C sniper rifles.

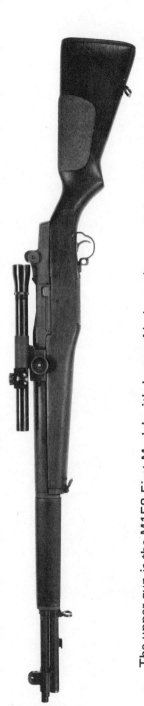
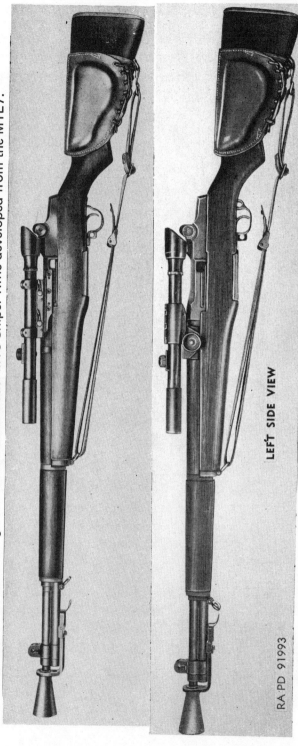

The upper gun is the M1E8 First Model with Lynan Alaskan telescope and an adjustable cork covered cheek rest. The second gun is the standard Model M1C Sniper Rifle developed from the M1E7.

LEFT SIDE VIEW

The lower gun is the M1D developed from the M1E8. It is shown with its accessories, a flash hider on the muzzle and a leather T4 cheek pad.

GARAND SNIPER MODELS

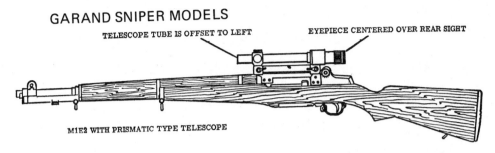

M1E2 WITH PRISMATIC TYPE TELESCOPE

In order to meet urgent overseas requirements for sniper rifles, tests were conducted by the Infantry Board to alter and fit several M1903 Springfields with commercial types of telescopes. As a result of the tests, the Weaver 330C was chosen to be our new sniper scope and designated the M73B1.

At the time Headquarters, Army Ground Forces, recommended the standardization of the M1903A4 Sniper Rifle, it also recommended that the M1 Rifle be equipped with a telescope.

Since the M1 clip loaded from the top, the gun presented a mounting problem. Naturally the method of feed could not be changed, so the only solution was either a prismatic type telescope offset to the left with the eyepiece centered, or a straight type telescope mounted to the left side of the rifle.

The designation M1E2 was assigned to the M1 Rifle with the prismatic type of telescope. The odd looking telescope and mount was developed for the Ordnance Dept by International Industries.

The eyepiece of the telescope was centered over the rear peep sight on the M1. It extended forward for a short distance and fitted into a small casting containing a right angle prism. The balance of the telescope tube extended from the other side of the prism casting. This gave the telescope the appearance of a crankshaft.

A quick detachable type rail mount was also developed and mounted to the side of the M1. It was an odd looking rig but it allowed the M1 to be loaded in a normal manner and allowed the empty clip to eject when the last round was fired.

Along with the prismatic telescoped M1, the Infantry Board tested an M1 with an offset Stith-mounted Weaver 330.

Springfield then fitted two (2) M1's with telescopes, the M1E7 and M1E8. The E7 utilized a Griffin and Howe telescope mount and a Lyman Alaskan (M73) telescope.

The drawback with the G & H mount was that it required the drilling of 5 holes in the M1 receiver. Three were tapped and two were tapered for dowel pins.

The E8 did not require any holes in the M1 receiver, but a steel block had to be pinned to the chamber end of the barrel for a mount base and the hand guard had to be shortened.

Both guns fulfilled the requirements fairly well. Neither design required any extensive modification of guns coming off the war production lines.

Of the two designs the Griffin & Howe mount proved to be the sturdiest and the one that required the least amount of critical machine time; but the mount on the E8 proved easier to mount and dismount.

The Lyman Alaskan telescope was issued with either crosshair or tapered post type reticles. To distinguish between them, the telescope with crosshairs was designated the M73 while the telescope with the post was called the M83.

A cork covered metal cheek pad was tried on the M1E8 and designated the T6. Tests at Aberdeen indicated the need for more face support. This was added and the cheek piece was revised to a TGE1. However, in the latter part of October 1944, it was recommended that this item be dropped.

On the basis of Infantry Board tests of the M1E7 and M1E8 Rifles, the M1E7 was standardized in June 1944 as the M1C Sniper Rifle. This relegated the Springfield M1903A4 to the status of Limited Standard Sniper Rifle.

Because of the tooling required, the M1E8 was not adopted until September 1944. Then it's designation was changed to the M1D Substitute Standard Sniper Rifle.

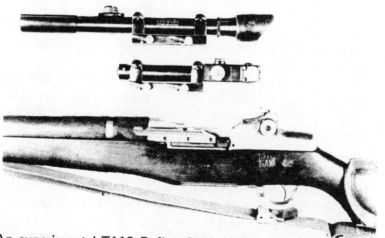

An experimental T110 Reflex Collimator sight was tested on an M1C Sniper Rifle. The small collimator sight was not accepted.

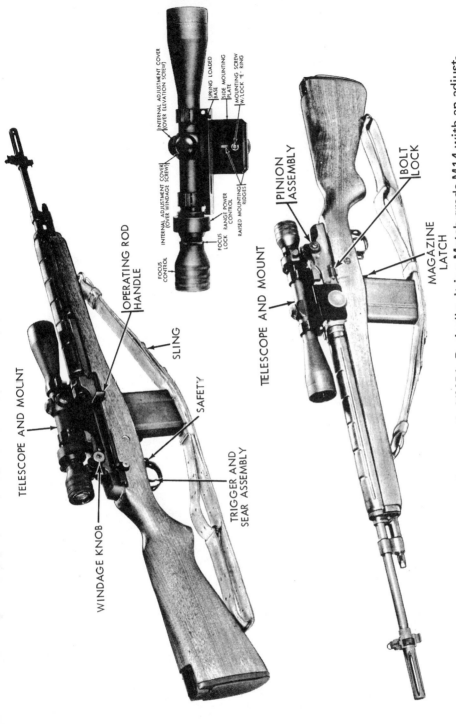

The latest U.S. sniper rifle is the semi-automatic XM21. Basically, it is a Match grade M14 with an adjustable ranging variable power (3X to 9X) telescope. Only Nato M118 National Match ammunition should be used with this weapon. The weight with equipment is 14 lb. 5-1/3 oz.

M1 GARAND PRODUCTION 1938-1945

M1 Garand rifle production began slowly at first. Self styled experts pointed out that some of the milling cuts in the receiver were difficult to cut and that the gun would be difficult to manufacture. Others lacked the foresight to see the advantages of the new semi automatic rifle over the bolt action.

1937 was no time to be spending money on a new gun. Chamberlain had even yet to say his famous "peace in our time"; depression was still rampant and war seemed far away. Springfield Armory began the production run and 100,000 guns later, Winchester was plugged into the pipe line. As the need for infantry weapons grew so did the list of suppliers. Harrington & Richardson, the old line maker of inexpensive rifles, shotguns and pistols was the only other arms manufacturer to make the M1 rifle. International Harvester, the only other maker of M1's slowed their line of farm machinery in order to produce several hundred thousands Garands.

All in all, over 4-1/4 million M1's were produced from 1937 to the end of WW II in 1945. Of this huge stock of M1's, many were supplied to friendly NATO or Seto countries, others sold to shooters thru the NRA or cut into pieces to demilitarize them.

During the 1950's and 1960's, as quickly as the Army was cutting the receiver for scrap, enterprising individuals were rewelding the receivers and remachining the welded area to make the gun sound again. These early ecologists turned scrap into gold until a bureaucratic regulation made real scrap out of the M1 receivers by crushing them beyond repair. The quality of the rewelds varied but because of the location of the non load bearing portion of the receiver usually cut, there was little chance of the welding affecting the locking lug area. A number of these rewelds were cut down and sold as so called "tanker" models.

A number of M1's were reimported into America from England after WWII. These early M1's were supplied to England under the Lend Lease Program before we entered the war officially. They can be easily identified by the broad arrow proof mark and "not English made" stamped on the barrel. Some have vestiges of a red band painted around the fore end to identify the gun as 30-06 caliber. This was done to prevent raw soldiers from trying to use 303 British ammo in a gun chambered for 30-06 caliber ammunition.

The rifle above is the Model T20, the first of the full and semi automatic fire M1 Garands. Springfield Armory delivered the first gun to Aberdeen Proving Grounds in November 1944.

The upper two photos show a right and left view of the T20E1. It incorporated a number of changes recommended after the T20 was tested at Aberdeen Proving Grounds. The lower gun shows a T20E2. It incorporated further changes from the E1 gun. Springfield built ten (10) E2's for extensive tests. The gun above is one of the first batch modified to fire the short T65 cartridge.

FULL AUTOMATIC GARANDS DURING AND AFTER WW II

The wide use of full automatic weapons by the Germans during WW II prompted us to develop a new family of automatic weapons. Early in 1944 the U.S. Ordnance Dept. directed Springfield Armory and the Remington Arms Co. to develop a new paratroop rifle having the following characteristics—weight 9 lbs. complete less magazine; overall length 26 inches with stock folded; bipod; capable of full and semi automatic fire; able to fire grenades and be close to the M1.

The Springfield rifle series was designated as the T20 and the Remington series the T22. The specifications were changed shortly and the folding stock was eliminated. By November 1944 Springfield delivered the first of the T20 modified Garands to Aberdeen Proving Grounds for tests. The gun fired semi automatic from a closed bolt for accuracy and full automatic from an open bolt. The bolt was kept open to cool the chamber and also to prevent a chambered round from a "cookoff" caused by a hot barrel. The first model T20 needed improvement. A muzzle brake prevented the gun from climbing during automatic fire but did not allow for mounting a bayonet, grenade launcher, or flash hider. Springfield's first T20's made use of a Browning Automatic Rifle (BAR) magazine. It proved too weak and it was suggested that a new stronger magazine be developed.

T20E1 incorporated the changes desired as a result of testing the T20. E1 rifles fired both semi auto and full auto from a closed bolt. Heat flow grooves were machined in the barrel to cool the chamber. The receiver was modified to take a telescope, night sight or grenade sight. The muzzle brake was modified to take a bayonet but it was still incapable of mounting a flash hider or a grenade launcher. A new box magazine and bipod was included in the E1 gun. During January of 1945 Aberdeen put the T20E1 thru an exhaustive series of tests. It worked well but had a number of failures to feed. This was remedied by hardening the breech end of the barrel. The gas cylinder was modified to permit easy attachment or removal of accessories and the hand guard redesigned to prevent charring the wood. After the tests, the gun was revised to incorporate the above ideas. This new gun was called the T20E2.

The T20E2 differed from the T20E1 mainly in the muzzle brake, bolt, and receiver. A redesigned muzzle brake permitted the attaching of a grenade launcher, flash hider and bayonet. A roller was attached to a lug on the bolt to improve the performance in the rain. Last, but not least, the receiver was lengthened slightly to allow the bolt to travel further to the rear. This improved the feeding during full automatic fire.

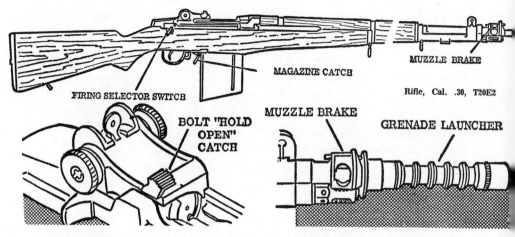

In order to get the guns to the combat zone, in May 1945, the Ordnance Committee recommended the T20E2 be designated as a Limited Procurement Type. Orders for 100,000 rifles were placed with Springfield. On August 14, 1945, hostilities ceased and the order was cancelled. Authorization was granted for production of 100 rifles. Ten T20E2's were completed in July of 1945 and seven (7) were sent to Aberdeen for testing. The balance of the order was completed and the guns converted from time to time to various new "T" numbers.

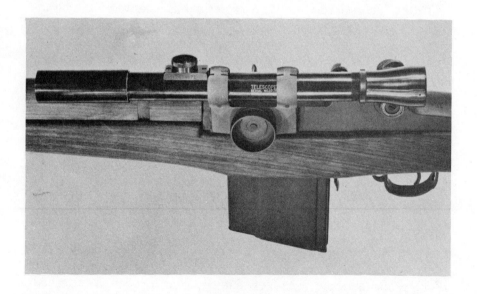

Telescope mounting rails were incorporated into the T20 receivers. A simple rugged single screw mount was used to hold a Lyman Alaskan telescope on the T20E2 rifle shown above.

REMINGTON GARAND CONVERSION TO FULL AUTOMATIC MODELS T23 AND T24

Strangely enough, the T23 and T24 Garands were designed before the T22. In order to check their preliminary designs, Remington Arms converted two M1 rifles to full and semi automatic fire. The first model designated Rifle Caliber 30, T23 used a selective fire control system based on an independent hammer release. The Rifle Cal 30, T24 achieved selective fire by means of an independent sear release.

Both models performed satisfactorily at the Remington test range before being turned over to the Ordnance Dept in Nov. of 1944. During the Army tests, both guns exhibited minor problems. The T23 fired full automatic from an open bolt, to cool the chamber. This feature caused some trouble and removal of some of the components eliminated this problem.

The T24 rifle was preferable to the T23 because of the time necessary to convert a standard M1 to a T24 was less than that of the T23.

The Ordnance Dept tabulated the results of their tests and passed them on to the Remington Arms Co. for incorporation into the T22 rifle.

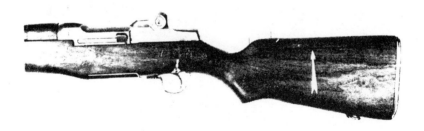

U. S. RIFLE, CAL. .30, T23

The gun shown above is the Remington T23 with an experimental high comb stock. The stock was tested to determine if it had any effect on the accuracy during full automatic fire. It apparently had little effect. The fire selector was a small button right over the trigger. It changed the gun from full auto fire to semi auto fire.

REMINGTON T22 RIFLES

The T22 rifle was basically the Remington T23 conversion of the M1 Garand, with the changes recommended by the Ordnance Dept incorporated into it. Remington went full speed ahead on the T22 rifle and submitted a sample rifle a short time later. The T22 rifles, after test, required only minor modifications to improve the functioning of the magazine release and bolt hold open. These modifications were included in the T22E1 rifle and based on successful tests at Aberdeen Proving Grounds, Remington was authorized to complete three models of the T22E2. The E2 model incorporated a change in the trigger group, magazine catch, gas cylinder, muzzle brake, and bipod. A study revealed that the design of the T22E2 rifle more readily lent itself to the remanufacture of the M1 Rifle because it used the standard M1 receiver rather than the longer receiver used by Springfield's T20E2 rifles.

The ending of WW II put a stop to Remington's work on the T22E2 but the Army kept Springfield busy making numerous variations on the basic M1 Garand.

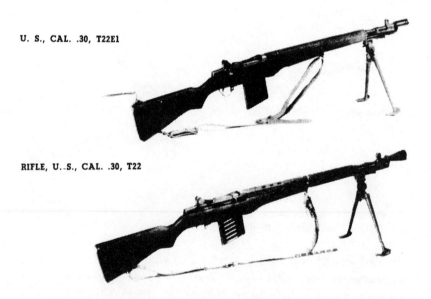

U. S., CAL. .30, T22E1

RIFLE, U. S., CAL. .30, T22

The ribbed magazine shown here on the T22 rifle was later changed to a BAR type magazine.

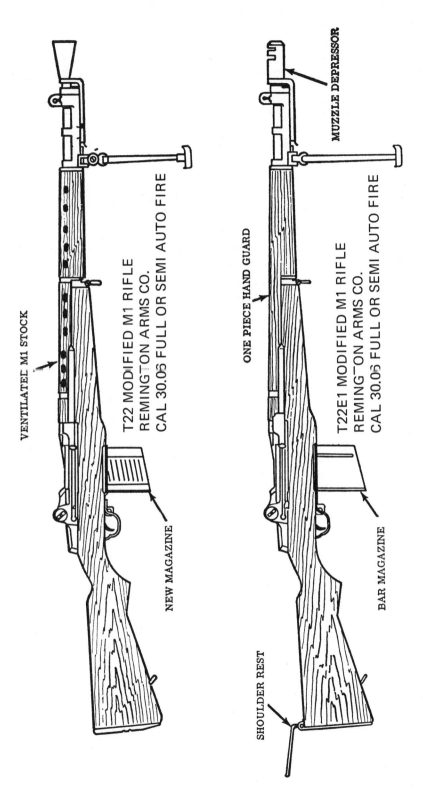

During the latter part of WW II, the Remington Arms Co. developed the T22 series of Modified M1 rifles. The design was in some ways preferable to the T20 series developed by Springfield Armory.

MAGNESIUM M1 STOCKS

The difficulty of obtaining black walnut wood for M1 stocks became apparent as production of the rifle increased. Birch and Cherry were authorized as substitutes but at one time it seemed that even these common hard woods would be in short supply.

In order to maintain M1 production, it was thought wise to investigate the possibility of using a light metal stock. So the Ordnance Dept contracted with the Light Metal Alloy Foundry at the State College of Washington. The foundry succeeded in sand casting several Magnesium alloy stock. They were machined a bit and had the surface finished before installing them on standard M1 barreled receivers. The guns shot well and there was no tendency to transmit noises or vibrations. The weight was about the same as the standard M1 stock but it was four times as strong. Eighty shots were fired as rapidly as the clips could be loaded but it caused no uncomfortable heating of the stock. Since the draconian shortage of hardwood did not materialize, the metal stock project was shelved.

STOCK, RIFLE, T9

TEFLON COATED GARAND

In an effort to devise a protective finish that would also act as a lubricant, a small quantity of M1's were coated with Dupont Teflon. Some of the test guns were left in the natural green color of the Teflon; others were finished in a gray blue color. The idea was dropped because the coating was not durable enough.

It was found that the coating process subjected some heat treated parts to a greater temperature than necessary.

It is interesting to note that a quantity of 30 caliber M2 ammunition was also teflon coated. It too fell by the wayside.

NATIONAL MATCH M1 RIFLE

In March of 1953 Springfield Armory was directed to furnish 800 selected M1 rifles for the newly reinstated National Matches at Camp Perry. At that time Springfield was still manufacturing M1's so it was an easy matter to select a batch of match quality guns.

From 1954 thru 1959 Springfield produced National Match quality rifles by two methods; one by doing minor gunsmithing operations on newly manufactured guns and second by rebuilding existing National Match M1 rifles.

In 1959 production of new M1 rifles was phased out. All subsequent issues of National Match rifles were made from rebuilt M1's.

As a result of a two day meeting held at Springfield Armory in Sept. of 1956, the Ordnance Dept. made funds available to conduct an engineering program to improve the accuracy of the National Match M1. The study resulted in a series of minor but important changes in thread fits and mechanical tolerances. By 1959 all National Match Garands had new rear sights with half minute of angle windage adjustments. Each sight part was marked NM.

The funds provided in 1957 as a result of the Sept. 1956 meeting allowed Springfield to develop new instruments. The straightness and bore diameter of a barrel could be measured far more accurately with the new instruments. This resulted in a tightening of the accuracy acceptance requirements for 1959

The following year by year breakdown indicates the quantity of National Match Cal. .30 M1 rifles produced by Springfield Armory:

Year	New	Rebuilt	Total
1953	800		800
1954	4184	499	4683
1955	3003	314	3317
1956	5050	550	5600
1957	4184	499	4683
1958	1295	731	2026
1959	2877	2652	5529
1960		8663	8663
1961		1410	1410
1962		4500	4500
1963		3639	3639

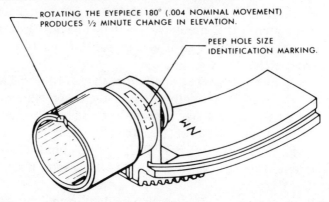

NOTCH INDICATES POSITION OF PEEP HOLE.
NOTCH AT TOP RAISES POINT OF IMPACT OF BULLET.
NOTCH AT BOTTOM LOWERS POINT OF IMPACT OF BULLET.

ROTATING THE EYEPIECE 180° (.004 NOMINAL MOVEMENT) PRODUCES ½ MINUTE CHANGE IN ELEVATION.

PEEP HOLE SIZE IDENTIFICATION MARKING.

The sight shown above was developed to allow a finer elevation adjustment than possible with the standard M1 sight.

National Match M1's. The allowable extreme spread was reduced to .20 at 100 yards.

Although the M1 Garand National Match rifles saw many years of service at Camp Perry, 1959 sounded the death knell for this fine rifle. For it was in 1959 that Springfield Armory set about to develop M14 National Match rifles. Even though the M14 (T44E4) had been classified as the standard rifle in June of 1957, development of the National Match M1 continued.

Although there were changes in the rear sight aperture, the most significant change to take place in the history of the NM Garand was the introduction of glass bedding. Even the best walnut stocks respond to moisture and temperature changes. These variations at time affected the accuracy of the M1 at critical receiver contact points. To aleviate some of these factors, National Match stocks were routed out at the points shown in the illustration.

A mixture of chopped fiberglass and epoxy resins were applied to the routed areas of the stock. The receiver and assembly were coated with a mold release agent and assembled to the stock. Excess epoxy was removed and the stock and receiver were cured for (8) hours. When completely cured or hardened, the stock was stamped with the last four digits of the receiver serial number to prevent any interchange.

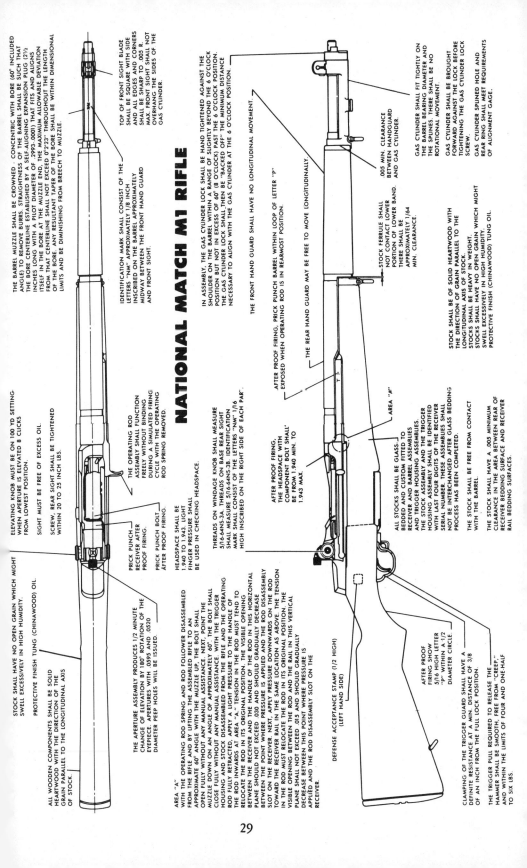

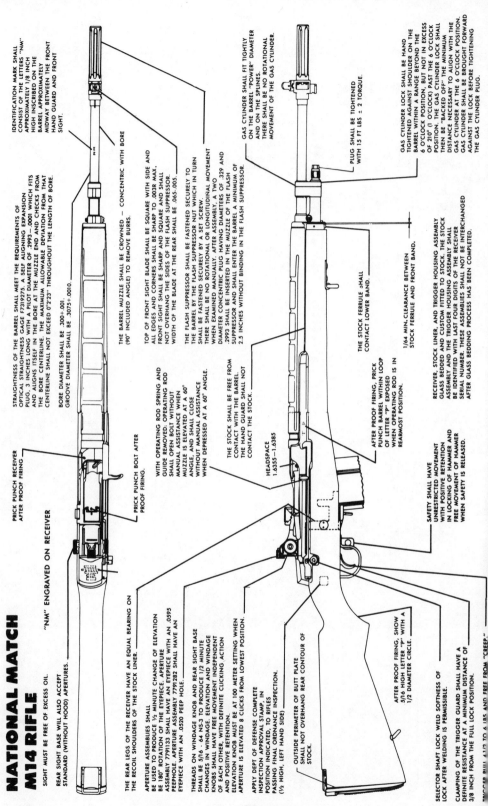

THE GARAND "T" SERIES

The M1 Garand was under constant development by the Ordnance Dept during WW II. Since the Garand was a peacetime development, many of the problems that turned up in combat had not been anticipated

Earlier in the book we saw a list of "E" changes that were added to the basic M1. It would be wrong to say that an "E" variation represents an attempt to solve some sort of trouble. The M1E7 & 8 for instance, was simply the prototype of the sniper models M1C & D. In some cases, the "E's" were experimental attempts to simplify the mechanism.

Interspersed between the experimental M1 "E" series, we find a number of test rifles with a "T" series number. The group that concerns us at this point is the batch of numbers from T20 to T44. Most of the rifles in this group are based on the Garand action, but there are a few that are in no way related to the Garand.

TABLE OF "T" NUMBER RIFLES

T20:
This rifle was the first of the selective fire M1's. It was developed at Springfield Armory during 1944 to Jan 1945. The gun weighed 10 lbs, had a 20 round box magazine and a 500 shot per minute rate of fire, length 43.61 inches.

T20E1:
A modification of the T20 to add minor design changes. (See page 21.)

T20E2:
A further modification of the T20 gun. This gun was developed enough to be designated for Limited Procurement in May of 1945. Weight 9.63, length 44.88 inches, rate of fire 700 rounds per minute.

T22:
Remington Arms Company developed this version of the M1 simultaneously with Springfield's T20. Weight 9.6 lbs, length 43.60 inches, rate of fire 500 rounds per minute. (See page 24.)

T22E1:
A modified version of the T22 with simplified trigger group and improved magazine catch. (See page 24.)

T22E2:
The E2 gun was a further simplification of the basic T22 rifle. The project was terminated in March of 1948.

T23:
Remington Arms Company's full and semi automatic version of the M1. (See pages 23-24.)
T24:
Remington Arms Company's full and semi automatic version of the M1. (See pages 23-24.)
T25:
A lightweight rifle designed by Springfield in 30 NATO caliber. (Not a Garand)
T26:
A carbine or short rifle version of the M1 developed from the M1E5. (See page 8.)
T27:
A modified M1 in (7.62mm) 30 NATO caliber. Capable of full or semi auto fire by installing an auxiliary device in the field. Weight 9.6 lbs, rate of fire 600 rounds per minute using standard 8 round Garand clips. Project terminated in March of 1948.
T28:
A lightweight rifle developed by Springfield with straight stock, full and semi auto fire in 30 Nato caliber. Development stopped in 1950. (Not a Garand.)
T31:
An odd lightweight rifle with the magazine behind the pistol grip and trigger. Full and semi auto in 30 Nato cal. (Garand design)
T33:
The Clarke rifle lightweight, full and semi auto fire in 30 NATO caliber. Project stopped in 1950. (Not a Garand.)
T34:
A Browning auto rifle converted to fire the 30 NATO cartridge.
T35:
This rifle was a modified M1 converted to fire the 30 NATO cartridge. It had a new barrel and a magazine insert to handle the shorter cartridge. Fifty were built but they did not perform well in rain and ejected empty cases into shooters face. Terminated in 1950. Weight 9.8 lbs, semi auto fire only using standard Garand clips.
T36:
This gun was a modified T20E2 rifle. It was converted to fire the 30 NATO cartridge. It fired full and semi auto from a closed bolt. Weight 8.43 lbs, rate of fire 600 rounds per minute, overall length 43 inches. (Suspended in 1950.)

The T35 was the first of the Army M1's to be converted to the T65E light-weight rifle cartridge. The two lower guns are right and left views of the T35—SH. This rifle had an integral side loading box magazine. Another T35 was altered to use a Johnson rotary type magazine.

T37:
A lightweight rifle modified from the T20E2 Garand incorporating features from the T36. It had a lightweight 22" barrel and lightweight stock and fired the 30 NATO cartridge. Development was suspended late in 1950. Weight 8.2 lbs, overall length 42.25", rate of fire 750 rounds per minute.

T38:
The T38 rifle was a modified version of the T35 to incorporate a rather unusual integral side loading magazine. Apparently the magazine could be loaded with the action closed and the magazine in place. The cocking lever was turned upward to prevent interference with loading the magazine.

T20 to T44:
The T44 rifle was the culmination of all the "T" series guns in search for a lightweight, selective fire infantry rifle to replace the M1 Garand. It was a "potpourri" of "T" rifles. It had the basic T20E2 Garand locking mechanism. The front end components of the T25 rifle, the magazine from the T31 and the gas system of a modified T37 with a gas expansion-cutoff system. It also incorporated a prong type flash supressor and an automatic pressure relief valve for launching grenades. It weighed 8.70 lbs and was 44.75 inches long. It fired the 30 NATO cartridges at a rate of 800 rounds per minute.

T44E1:
In an effort to replace the heavy Browning Auto Rifle (BAR), a heavy barrel T44 was fabricated in Oct. 1951. While it had the same action, the trigger mechanism was redesigned to incorporate a rate of fire reducer. It had a low rate of 550 rounds per minute and a high rate of 735 RPM. With a heavy barrel, heavy stock, and a bipod, the gun weighed 11.5 lbs.

T44E2:
This rifle was not a mere modification of the T44 rifle. It was a drastic design change. The gas expansion system was dropped and a gas impingement system substituted. A new operating rod, new bolt, trigger housing and a reduced volume grenade launcher were also incorporated into the design. Weight 8.2 lbs and length 42.25" overall. Rate of fire was 700 rounds per minute.

T20E2 rifles were being constantly modified. The T37 lightweight rifle used a T20E2 receiver modif' to fire the T65E3 (7.62 Nato) cartridge. The lower gun is an early version of the T44 rifle. Notice th the bayonet latch is in the middle of the bayonet gr p rather than at the top of the grip.

T44E3:
The T44E3 was a heavy barrel version of the T44E2. It resembled the heavy barreled E44E1 but used the gas impingement operating system instead of the gas expansion system. The impingement system proved unreliable and the gun was dropped.

T44E4:
This is the rifle that eventually became the M14. Basically, it was the T44 rifle with a gas expansion-cutoff system and a number of minor changes. In June of 1957 the T44E4 rifle was classified standard as the M14.

T44E5:
This heavy barrel version eventually became the M15 rifle. It too used a gas expansion-cutoff system and had the same trigger mechanism as the original T44. It was dropped in Dec. of 1959 after the M14 proved as accurate in tests equipped with a bipod.

T44E6:
The E6 rifle is probably the ultimate in the Garand cycle. When the standards were set for the lightweight rifle back in 1945, they called for a 7 pound weight limit. The 8.7 lb M14 while lighter than the 9.7 lb M1 was still far off the mark. Springfield set about to build the lightest version of the M14. First of all, the barrel was cut to 20" and reduced in diameter. The flash hider was shortened and the bayonet lug removed. The wings were removed from the rear sight and cuts made in the receiver and gas cylinder plug. The manually operated gas valve for launching rifle grenades was removed. An aluminum butt plate and magazine replaced the steel components and the stock was lightened. The wood stock was later replaced by a plastic stock. With all this work, the best they could do is cut the weight down to 7.8 lbs.

H & R designed a special M-14 specifically for guerrilla warfare.

While testing the lightweight T44 a few were modified to the T44E1. The lower picture shows the E1 with its heavy barrel, bipod and odd muzzle break and flash supressor. The upper gun is the final heavy barrel T44. It was designated T44E5 in October 1954. Later it became the M15 rifle.

The final acceptance of the T44 (M14) as our standard military rifle was not without a great deal of controversy. Development cost over 100 million dollars and twenty years. Scores of newspaper and magazine articles called the gun a failure and a waste of money. Whether the M14 was a failure or not is a matter of conjecture at this point since it has been largely superseded by the M16. The T44's development path ran into a number of road blocks. It was tested against the T25 and T28 rifles and seems to have emerged the winner, but the most serious challenge came in the form of the FN rifle built in Belgium by Fabrique Nationale d'Armes de Guerre. The FN rifle performed so well that Harrington & Richardson of Worcester,Mass. was given a contract for 500 of the Belgium rifles in order to evaluate it for production in the U.S. Apparently Hi Standard made a few of the T48 rifles as it was designated by the Army. Additional tests were then conducted. The T44 was matched against the U.S. T47 and the U.S. made FN rifle, the T48. The T47 was quickly eliminated and in further tests, the T44 was chosen over the T48. The reasons for choosing the T44 over the T48 were that the T44 was a pound lighter, better suited to mass production and training. A heavy barrel FN rifle, the T48E1 was also tested against the heavy barrel T44E5 with the same results. The T44E5 was chosen and called the M15. It was then declared obsolete in Dec. of 1959.

A number of sheet metal and bent wire magazine fillers or chargers were developed during the T44 program. Above, a lightweight T44, magazine is being loaded thru the receiver with (20) 7.62 Nato cartridges.

The two rifles shown here competed strenuously to become the new U.S. Infantry rifle during the 1950's. The upper gun is an early T44 -- the lower gun is a T48, one of the U.S. manufactured versions of the Belgian FN rifle.

The experimental rifles shown above were developed to replace the M1 but all failed for various reasons. See page 31. The upper gun is the T25; the center is the T33. The lower gun, the T47 came closest to replacing the Garand system.

In 1955 the Mathewson Tool Company converted a small quantity of T20E2 rifles to a gas impingement system. The right and left view above shows this unusual rifle. The lower is an early model T44 (M14) with the wood hand guard and slab side receiver.

As we have seen early in the book, the search for a full and semi automatic rifle to replace the battle tested M1 Garand was long and tedious. Numerous designs, revisions and operating systems were tried, but when the dust settled, the Army adopted the T44 in May of 1957. The T Number was dropped and the number M14 was adopted for the regular weight barrel and M15 for the heavyweight barreled version. Tests soon showed that there was no great advantage in the M15 version and it was dropped in December 1959. When the M1 and M14 are examined together, the lineage is obvious but the M14 contains a number of changes. The obvious changes are the 20 round detachable magazine, the roller on the bolt to reduce friction, the new gas system and the automatic fire selector. Later the M14E1 was produced with a folding stock. In 1968, the M1E2 became the M14A1. It differed from the M14 in that it had a straight line stock with a pistol grip and a folding front hand grip. The A1 version has a hinged shoulder rest and a rubber recoil pad. Last, but not least, there is a bipod and perforated steel sleeve stabilizer that slides over the flash suppressor. M14 production was halted in 1963.

To the right, the MI4 equipped with the Starlight Scope (image intensifier) for night sniping.

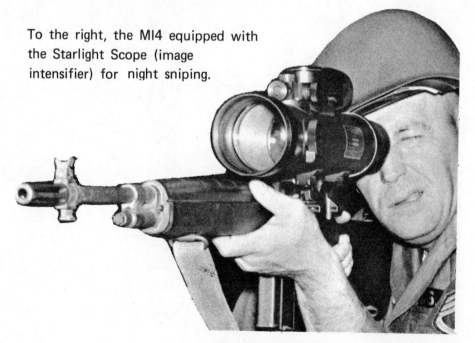

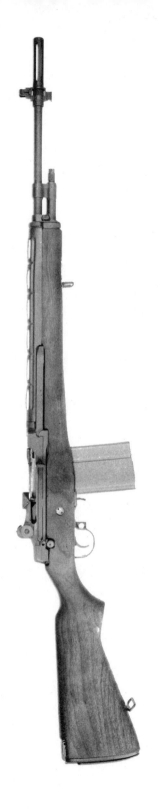

The upper rifle is a late issue National Match M14. Note that the selector switch has been replaced with a non-functioning knob. The lower gun is the rare T44E6 Super Light Weight rifle, designed to bring the T44 down to 7 pounds.

In an effort to make the M14 more compact for the paratroops, a number of folding stocks were developed by Springfield. The upper two photos show the Type III stock open and closed. The lower photo shows the M14 with the Type V side folding stock and folding pistol grip.

Straight line stocks were added to the M14 to help prevent muzzle climb. The upper gun is the M14 (USA1B) but the second, an M14E2 and the lower gun is an M14 with a Type V stock in open position.

NAVY M1 CONVERSION FROM 30 CAL M1 TO 7.62 NATO

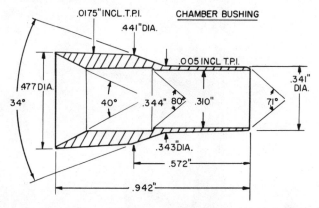

The Navy has little need for an infantry type weapon aside from the Marines. So in an effort to save funds, in 1963 the Navy designed a spacer or bushing that decreased the length of the M1 chamber. This spacer allowed the M1 to fire the shorter 7.62 NATO cartridge. The spacer was developed for the Navy by H.P. White Labs and a quantity of M1's were converted for the Navy by the American Machine & Foundry Co. of York, Pa.

The final change consisted of coating the bushing with Loktite, inserting it into the chamber and seating it by firing several rounds. The gas port was enlarged to .1065 and a plastic spacer added to prevent loading a clip of 30.06 cartridges. All the Navy converted M1's are engraved on the left side of the receiver, just below the sight, with the legend 7.62 NATO. The letters are about 1/4" high and filled with white paint.

THE FOLLOWING PAGES (47-65) WERE REPRODUCED FROM AN EARLY ORDNANCE REPAIR MANUAL. IT INDICATES THE PROCEDURE FOR REVISING THE EARLY PRE—WAR M1'S INTO THE REVISED WARTIME GARANDS.

14. MARKING OF REBUILT WEAPONS. *a. Initials.* All machine guns, pistols, rifles (including automatic rifles) and carbines rebuilt will be stamped with the initials of the rebuilding establishment in the United States; weapons rebuilt by oversea base shops will not be stamped. Initials identifying the establishment rebuilding a machine gun are stamped on the right hand side plate directly to the left of the serial number. On pistols and Browning automatic rifles the initials will be stamped as close as possible to the serial number. All rifles (except the Browning automatic rifle) and carbines will be stamped on the left side of the stock between the hand grip and the butt plate. If the weapon is subsequently rebuilt at another establishment, the new identifying initials will be placed directly below those preceding. If the weapon is rebuilt at the same establishment as before, new initials need not be added. The establishments and the initials to be used are as follows:

Augusta Arsenal	AA
Benicia Arsenal	BA
Mt. Rainier Ordnance Depot	MR
Raritan Arsenal	RA
Red River Arsenal	RRA
Rock Island Arsenal	RIA
Springfield Armory	SA

b. Proofmark. Rifles (except Browning automatic rifles) and carbines will be marked after proof firing with a "P" on the under side of the stock immediately in the rear of the trigger guard to show that they have been proof fired. The "P" will be stamped on the under side of the forearm on Browning automatic rifles. (See requirements for proof firing.) If the rifle or carbine is again proof fired and the first proofmark is still visible on the stock, the next proofmark will be applied in line with the original proofmark and toward the butt plate.

CAL. 30 U. S. RIFLES M1 AND M1C. *a. Slight assemblies.*
(1) *Front sight.*
    (a) Blade will be straight and top square.
    (b) Check for looseness.
    (c) Wings will not be burred or bent.
    (d) Where facilities permit, the rifle will be targeted. It may be boresighted if prior approval of the boresighting method is obtained from Chief of Ordnance, Washington 25, D. C., ATTENTION: ORDFM. Targeting

will be done at a range of at least 25 yards. When rifle is targeted or boresighted, there is no requirement in final inspection for checking the height of the front sight blade. Otherwise, the blade will not be less than 0.718 inch from top of blade to bottom of base. After targeting, the front sight will not overhang the base. (On sights of early manufacture which are as wide as the base, the front sight may overhang the base by a dimension which is the equivalent of five clicks.)
  (e) Check to assure that screw is tight.
  (f) Blade will not "shine."
  (g) Front sight screw seals will not be replaced.
(2) *Rear sight.*
  (a) Windage and elevating knobs will operate with distinct clicks and without binding.
  (b) Cover will fit securely.
  (c) With 100-yard elevating knob graduation opposite index line on the receiver, it will be possible to depress the aperture from 1 to 15 clicks after targeting.
  (d) Aperture will be under vertical spring tension from the cover (assembled cover should not be peened to secure tension).
  (e) Aperture will not have over 0.013-inch horizontal

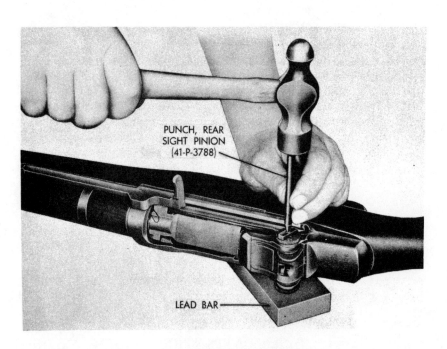

Staking rear sight pinion.

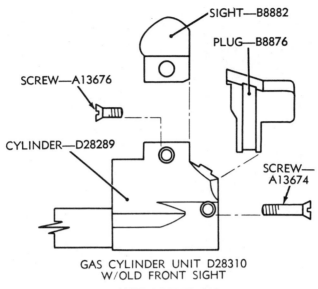

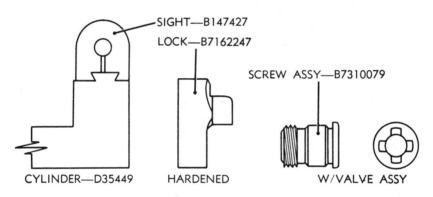

Gas cylinder units.

movement in the base at 600 yards elevation. (No gage required.)

(f) Check to assure that end of pinion is staked as shown in figure 39 and as prescribed in TM 9-1275. Place a lead bar or a piece of hardwood under the elevating knob before staking so as not to damage rear sight assembly. Pinions C113697 having minor cracks due to staking need not be replaced.

(*g*) Locking pin (3/16 x 0.075 in. drill rod) will be installed on the telescope mount assembly of the M1C to keep front and rear telescope support screws from working loose.

(*h*) Check to assure that rear sight elevating knob screw is tight.

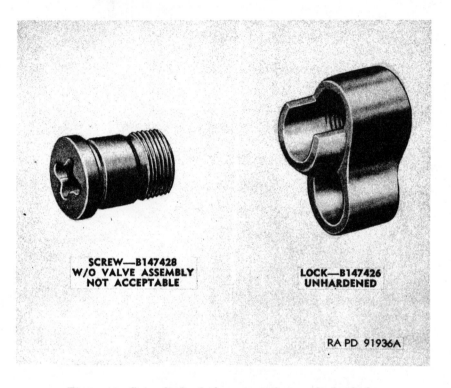

Figure 41. Gas cylinder lock screw and gas cylinder lock.

b. *Gas cylinder group.*
   (1) No dents or burs will be apparent.
   (2) Gas cylinder port will be clean.
   (3) Check to assure that the gas cylinder has a satisfactory dichromate black finish.
   (4) Rifles having screw type gas cylinders are unserviceable and irreparable (fig. 40).
   (5) All rifles will be equipped with gas cylinder lock screw with valve assembly B7310079 (fig. 40). Visually inspect assembly for cracked valve and to assure that it is clean. Check functioning by hand.

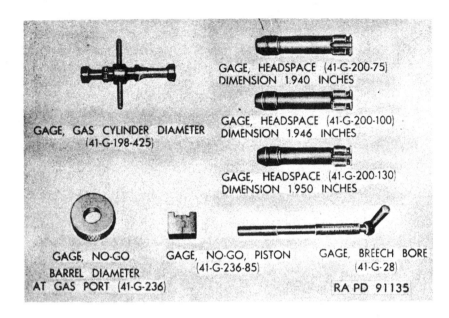

Figure 42. Gages for inspection.

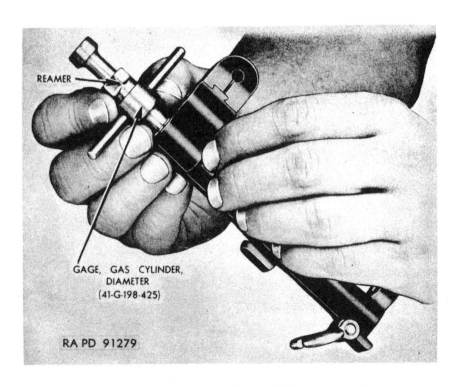

Figure 43. Gaging bore of gas cylinder at gas port.

(6) Gas cylinder lock should be alined with gas cylinder by screwing it down snugly and then backing it up into alinement. Gas cylinder should then be tapped forward against the lock. (Refer to TM 9-1275.) If alinement of gas cylinder lock and gas cylinder is found difficult, selective assembly will be used.

(7) Hardened gas cylinder lock B7162247 is preferred and lock B147426 is usable (fig. 41). However, the present stock of the hardened lock (B7162247) is being conserved for use with the grenade launcher M7A1 where its use is essential.

(8) Check inside diameter of gas cylinder with gas cylinder diameter gage 41-G-198-425 (figs. 42 and 43).

(9) Overall rotational movement at the top of the front sight blade when assembled to the barrel will not exceed 0.015 inch. Check the looseness of the gas cylinder with gage 41-G-98-820. This dimension will be maintained by selective assembly of the gas cylinder. Any gas cylinder which allows excessive motion will be discarded as unserviceable. (A gas cylinder will be tried on several barrels before it is discarded.)

(10) The gas cylinders with a diagonal saw cut will be used,

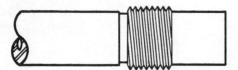

BARREL D28286
SHORT BARREL W/SINGLE SPLINE
NOT ACCEPTABLE

BARREL D35448
LONG BARREL W/TRIPLE SPLINE
ACCEPTABLE

Barrels—short and long.

if otherwise serviceable. Those with the straight saw cut will not be reused.

(11) Visually inspect bayonet stud, gas cylinder lock screw and front end of gas cylinder lock to assure against deformations that would interfere with assembly of bayonet or grenade launcher.

(12) Visually inspect to assure that stacking swivel and screw are in good condition and that screw is staked.

c. *Barrel and receiver group.*

(1) See that short barrel D28286 is replaced with long barrel D35448 (fig. 44).

(2) Head space will be as follows:

  (a) The component bolt will close on the 1.940-inch head space gage 41–G–200–75 (fig. 42).

  (b) The maximum limit, using the component bolt, will be 1.946 inches. This means that if the bolt closes without perceptible bite on head space gage 41–G–200–100 (fig. 42), the rifle is not acceptable.

  (c) The field test bolt 41–B–1587 will close on the 1.940-inch head space gage. This is necessary to assure interchangeability.

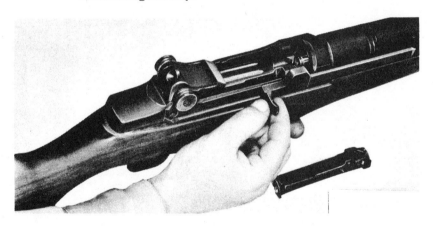

Checking head space with field test bolt and head space gage.

  (d) When the rifle is rebarreled every effort should be made to hold the head space as close to 1.940 inches as possible using the field test bolt. Under no circumstances will the head space exceed 1.946 inches using head space gage 41–G–200–100 with the field test bolt (fig. 45).

    Note. When using the head space gage, the bolt will be disengaged from the operating rod assembly. The rim of the gage will be carefully placed under the extractor against the face of the bolt if the extractor is assembled, thereby avoiding snapping of the extractor over the gage. In closing the bolt to check the depth of the chamber only the lightest finger pressure will be exerted. Further precautions shall be taken to assure that the hammer does not exert pressure against the bolt.

(3) Maximum breech bore reading is 0.305 inch, using breech bore gage 41-G-28 (fig. 42). Muzzle bore gaging is not required. Check diameter of barrel at gas port with

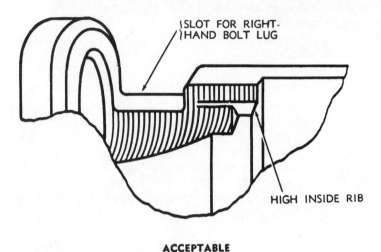

ACCEPTABLE

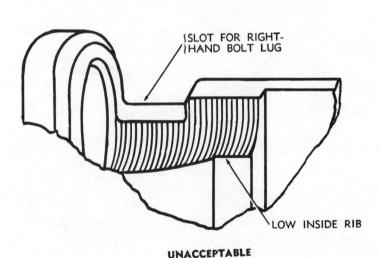

UNACCEPTABLE

Receiver D28291.

piston no-go gage 41–G–236 (fig. 42). No-go dimension is 0.5995 inch. Necessary precautions will be taken to assure that the barrel in each rebarreled rifle is alined properly with respect to draw in receiver.

(4) Check to assure that gas port is clean. Pass drill through hole to accomplish this. The hole diameter is 0.0790 inch + 0.0015 inch (drawing dimension).

(5) Pits on the receiver are permissible if they do not affect functioning.

(6) Receivers will have high inside rib (fig. 46).

(7) Receivers will be inspected under "black light" for cracks at the locking slot on the right-hand side of the receiver and at the bridge for the firing pin retracting cam. Another critical area that will be free from cracks or distortions is the rear end of the receiver where it is subjected to the repeated hammering of the bolt.

(8) Remove the mount bracket on the rifle M1C prior to refinishing and inspection to determine whether the mating surfaces of the bracket and receiver assembly were phosphate finished as an assembly and have any visible sign of corrosion.

Note. The mount bracket must be reassembled to the same receiver from which it was removed.

(9) Visually inspect receiver all over to assure against deformations or breaks at any thin sections, such as at the trigger housing locking cams.

(10) If, in rebarreling the rifle, it is found that the new barrel can be turned past the alinement position by hand, receiver gage 41–G–263–250 should be used to check the thread qualification of the receiver. Any appreciable quantity of receivers to which barrels cannot be assembled and which do not pass the gage test should be held in the shop and reported to the Chief of Ordnance, Washington 25, D. C., ATTENTION: ORDFM.

(11) Precautions will be taken in rebarreling to assure that the proper draw is maintained. If necessary, selective assembly should be used. Refer to TM 9–1275.

(12) Periodically check alinement of barrel and receiver as a check on assembling fixture. Use alinement gage 41–G–13–250 or similar gage which will accomplish the same purpose.

(13) Check to assure that only long fork follower rods (C64331) or later designs are used. Check visually for straightness. Check by hand to assure that rivets are tight. Scrap all *short* fork rods (fig. 47). Visually inspect to assure that rods are not deformed.

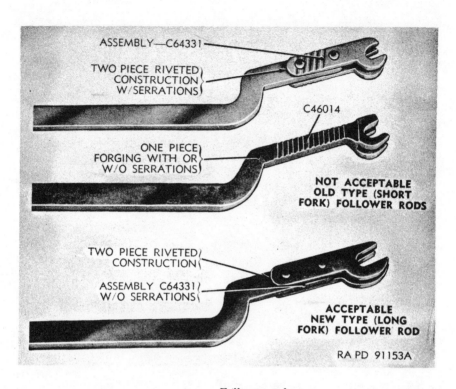

Follower rods.

(14) Visually inspect follower arm to assure that it is straight, that the clevis end is not deformed and that the four pivot points are in good condition.

(15) All followers C46004 will be of the latest type, which can be identified by the Revision No. 8 or above stamped on the under side of the follower tip or by guide arms at an angle (fig. 48). Care should be taken to avoid discarding satisfactory followers as some of these numbers were stamped very lightly.

(16) Visually inspect and check functioning of follower assembly by hand. If not necessary to disassemble, it may be refinished as a unit.

(17) Diameter of operating rod pistons should be checked with piston no-go gage 41–G–236–85 (fig. 42). Minimum piston diameter 0.525 inch. (These gages were ordered returned to Springfield Armory because incorrect amount of wear on piston was permitted. Gages modified to no-go dimension of 0.525 inch are satisfactory). Check height of engaging lug at rear end with a micrometer. It will not be less than 0.310 inch. Visually inspect to assure against battered or deformed lugs. Also, visually

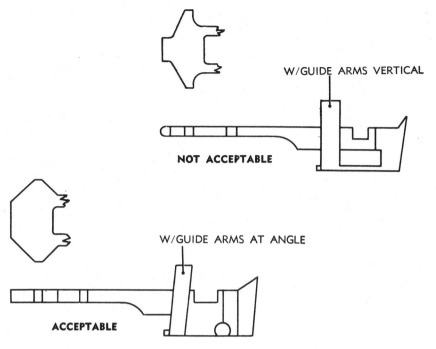

Follower C46004—showing correct angle of guide arms.

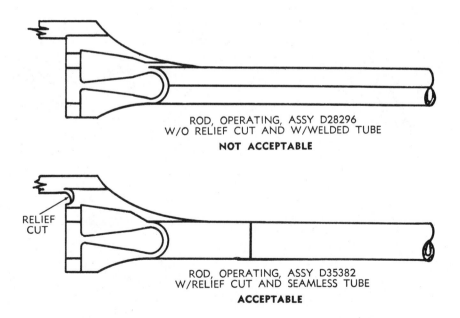

Operating rod modification.

inspect to assure that bolt operating cam is not excessively battered and that catch hooks are not excessively worn. Operating rods worn so that the lug on the rear end becomes disengaged from the guideway in the receiver during firing will be replaced.

(18) Check to assure that all operating rod handles have a relief cut radius at the junction of the arm and hook section of the handle (fig. 49). Those that do not have a radius will be reworked as prescribed in TM 9–1275.

(19) Inspect all operating rods under "black light" to assure against cracks around the cam area and at the forward end of the arm of the handle.

(20) Operating rod spring will be visually inspected to assure against mutilations or bent or broken coils. Free length should not be less than 19 inches. Weighing is not required.

(21) Lower band will be staked at both sides to retain the lower band pin securely. Slight shortening of the lower band pin (not les sthan 0.490 in. in length) is permitted, if necessary to assure adequate staking.

(22) Disassembly of accelerator from operating rod catch for inspection or refinishing is not required. Check accelerator to assure that it moves freely and is properly assembled. Check to assure that pin is staked at each end to retain it securely.

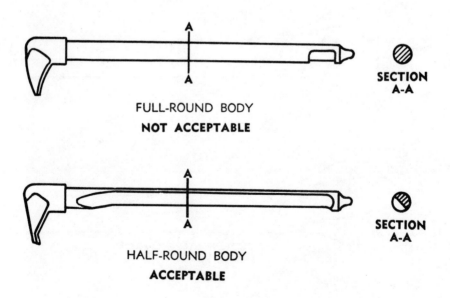

Firing pin B8879.

(23) Check accelerator cam on the bullet guide for height. This should be done with a micrometer or comparator. Dimension from cam surface to opposite side should not be less than 0.175 inch. Two designs were manufactured, one fabricated (brazed) and the other forged and machined; both are usable.

(24) Check each action for free movement of operating rod after assembly with bolt (without spring) as prescribed in TM 9-1275. Rod will not bind on sides of the lower band.

(25) The assembled weapon will be so timed that the operating rod catch will not disengage from the operating rod until the clip has been engaged by the clip latch. Check timing with gage 41-G-428-325.

d. *Bolt group.*
(1) Old style firing pins with full-round bodies will be replaced (fig. 50). In the future none of these firing pins should be sent to Springfield Armory for reworking.
(2) Check each bolt to assure that *no* bolts marked O-16 at the top rear end (in the location of the piecemark) are

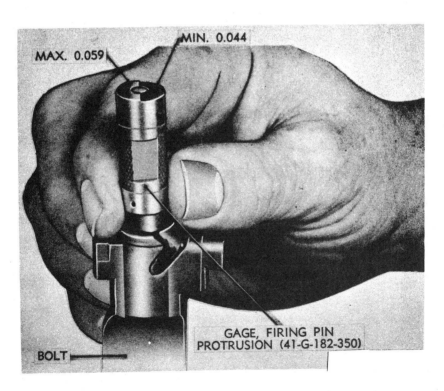

Gaging protrusion of firing pin.

reused. Any bolts with this mark (O-16) will be scrapped.

(3) Inspect each bolt for cracks under "black light" around the right-hand locking lug. Scrap any bolts that show evidence of cracks.
(4) Visually check to assure that hammer cocking cam is in good condition.
(5) Visually inspect the firing pin hole at face of bolt. The corner will be free from burs, but not rounded or chamfered.
(6) Visually inspect to assure that there are no burred edges at the two cams on the operating lug.
(7) Test firing pin for freedom of movement.
(8) Firing pin protrusion will be 0.044 inch to 0.059 inch using firing pin protrusion gage 41-G-182-351. Use gage 41-G-182-350 as a substitute (fig. 51).
(9) Inspect tension on extractor and ejector springs.
(10) All extractors will be replaced.
(11) Check assembled bolt by hand to assure free movement of firing pin and proper functioning of the ejector and extractor.

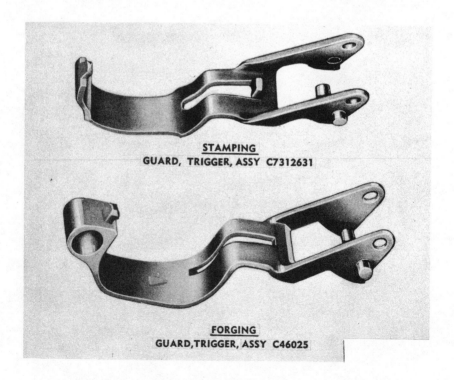

Trigger guards.

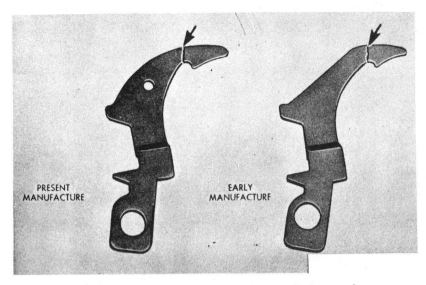

Safeties—important points to be inspected.

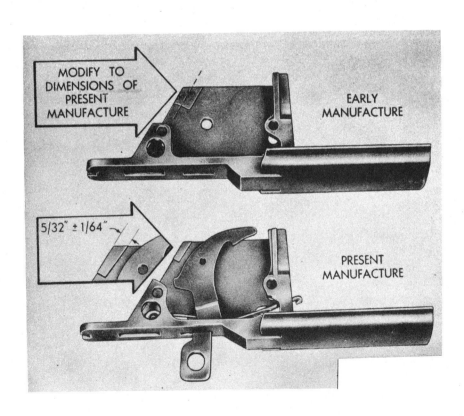

Modification of trigger housing.

e. Trigger housing group.
  (1) All components will be individually inspected visually.
  (2) Visually inspect trigger guard to assure that the locking lugs that engage receiver and hook at rear end are in good condition.
  (3) The forged C46025) and the stamped (C7312631) type trigger guards are usable
  (4) Either new or old type safety is usable (fig. 53).
  (5) The locking lug of the rear of the trigger guard assembly will have positive engagement with the recess provided in the trigger housing.
  (6) The clip ejector will be sufficiently strong to properly eject clip during function firing.
  (7) Trigger pull limits are as follows:

M1	M1C
Maximum—7½ pounds	Maximum—6½ pounds
Minimum—5½ pounds	Minimum—4½ pounds

  (8) Excessive force will not be required to close the trigger guard.

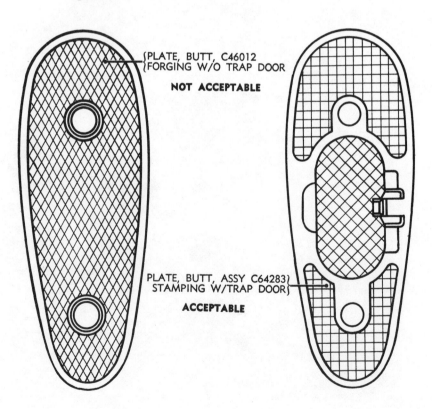

Butt plates.

(9) Pads on upper rear corner of trigger housings of early manufacture will be modified in accordance with figure 54.

(10) The information contained in TM 9-1275 pertinent to the Trigger Housing Group should be closely followed.

f. Stock and hand guard group.

(1) *Stock.*

(*a*) Butt plates will be of the type with a trap door, such as C64283 and D35466

(*b*) Inspect for cracks and mutilations. Scrap those stocks which cannot be reconditioned.

(*c*) Remove all metal parts and refinish any parts, if necessary.

> *Note.* In assembling or disassembling the stock ferrule the "band" should be loosened sufficiently to avoid stripping the mating ribs on the stock.

(*d*) If necessary to pad the contact surface for the trigger housing, mill out, block, and recut.

(*e*) Inspect visually to assure that there is sufficient clearance at the rear end of the trigger group notch for installation of the trigger group.

(*f*) Check to assure that the operating rod does not bind at the operating rod cut in the stock. Also check to insure that the lower band is in alinement with the operating rod.

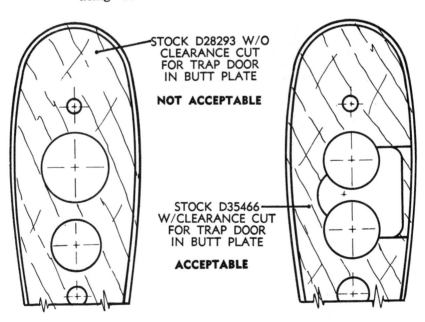

Stock modification.

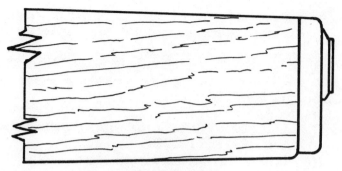

GUARD, HAND, FRONT, ASSY
C46028 W/EXTERIOR LIP

NOT ACCEPTABLE

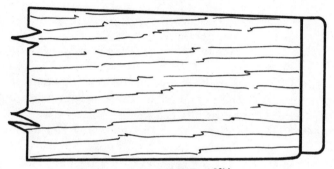

GUARD, HAND, FRONT, ASSY
C64245 W/O EXTERIOR LIP

ACCEPTABLE

Front hand guard assemblies.

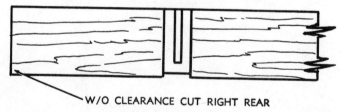

W/O CLEARANCE CUT RIGHT REAR

NOT ACCEPTABLE

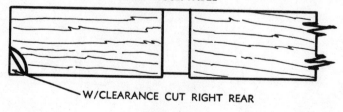

W/CLEARANCE CUT RIGHT REAR

ACCEPTABLE

Rear hand guard assembly C46024.

(g) Assure that the butt plate is fully seated and that wood does not protrude beyond the metal of the butt plate.
(h) Check for sufficient clearance for the operation of the butt plate. Modify the stock if clearance is insufficient (fig. 56).
(i) The fit of the stock should be such that the trigger guard will come to rest approximately 30 degrees from its locked position without forcing. If it is necessary to remove wood from the under side of the stock along the bearing surface for the trigger housing, extreme care will be exercised to preserve the 10-degree angle and to remove the same amount of wood from both sides of the stock.
(j) There should be no perceptible play between the action and the stock when the rifle is assembled.

(2) *Hand guards.*
(a) Front hand guard assemblies (C46038) with exterior lips are not acceptable
(b) Inspect for cracks and mutilations. Scrap those components which cannot be reconditioned.
(c) Remove all metal parts. Refinish these parts if necessary.
(d) Hand guard liner must be straight.
(e) Rear hand guard (C46024) will have clearance cut at right rear corner (fig. 58).
(f) Rear hand guard will fit securely.
(g) A certain degree of looseness in the front hand guard is permissible unless there is danger of the hand guard becoming disengaged from the weapon.

g. *Function firing.* Each rebuilt weapon will be function fired with 24 consecutive rounds without malfunction.

h. *Proof firing.* Weapons that have been rebarreled will be proof fired with one round of high pressure test ammunition.

TABLE OF M1 AND M14 SERIAL NUMBERS

U. S. RIFLE CALIBER .30 M1

CONTRACTOR	SERIAL NUMBERS		
Springfield Armory	1	TO	100000
Winchester	100001	TO	165000
Springfield Armory	165501	TO	1200000
Winchester	1200001	TO	1357473
Springfield Armory	1357474	TO	2305849
Winchester	2305850	TO	2655982
Unknown	2655983	TO	4100000
Field Service Depots	X4100001	TO	X4200000
Springfield Armory	4200001	TO	4399999
International Harvester	4400000	TO	4660000
Harrington & Richardson	4660001	TO	4800000
Not Assigned	4800001	TO	4999999
Springfield Armory	5000001	TO	5000500
International Harvester	5000501	TO	5278245
Springfield Armory	5278246	TO	5488246
Harrington & Richardson	5488247	TO	5793847
Springfield Armory	5793848	TO	6034228
Springfield Armory	UNKNOWN	TO	6099905

TOTAL PRODUCTION 1937 – 1945   4028395

U.S. RIFLE M14 CALIBER 7.62MM NATO

Fiscal Year	Springfield Armory	Harrington & Richardson	Winchester	Thompson Ramo Woolridge Inc.
1958	15600			
1959		35000	35000	
1960	32000	70082	81500	
1961	70500	133000		
1962	49000	224500	90000	100000
1963 & 64		75000	150001	219163
TOTAL	167100	537582	356501	319163

OVERALL TOTAL: 1380346

GARAND COPIES

The BM-59 rifle made by P. Beretta of Gordoni in Italy is based upon the design of the U.S. Garand .30 M1 rifle and resembles the U.S. 7.62mm M14 rifle. The BM-59 is gas operated selective fire, and is fed from 20-round detachable box magazines. There are several distinct models of the BM-59, differing primarily in stock design and muzzle device. All are mechanically identical. The Italian Army used the BM-59 Mark ITAL and the BM-59 Mark ITAL-A as standard weapons. The other versions are offered for sale and may be encountered in use in some of the small armies. BM-59 rifles were also produced in Indonesia. All versions of the BM-59 fire the 7.62x51-mm NATO cartridge.

The BM-59 can be used to launch 22-mm inside diameter tubed grenades. The compensator serves as a grenade launcher, as a muzzle brake and as a compensator. BM-59's used to fire grenades are equipped with a folding grenade sight.

M-1 MODIFICATIONS

Now you can have your M-1 Garand modified into a sporty carbine or fashioned into a modern M-14 by the leading expert in this field — Juan Erquiaga.

Erquiaga TANKER — Carbine version of the M-1. 18¼" barrel (six inches shorter than standard Garand). Retains all the features of the full-size M-1 yet only nine pounds fully loaded.
Modification Cost: $40.00. (For additional $10.00 will modify for 7.62mm NATO.)

Erquiaga EMFA-62 Modification — M-14 Version
Improved gas system . . . chambered for 7.62mm NATO cartridge . . . special muzzle brake . . . magazine fed, 20-round capacity.
Modification Cost: $65.00 (additional magazines available at $8.00 each).

A number of firms in the 1950's rewelded and rebuilt M1 Garands for U.S. shooters and collectors. Others, like Juan Erquiaga went one step further and brought out their own modifications of the basic M1 Garand.

The Japanese made copies of numerous foreign semi auto rifles. The lower gun is a Japanese version of the M1 Garand. The Japs made one great improvement in the feed. They used a Mauser type feed instead of the Garand clip. In this way, the gun could be loaded with loose cartridges. Caliber was 7.7mm and fired the standard 7.7mm Japanese rifle cartridge.

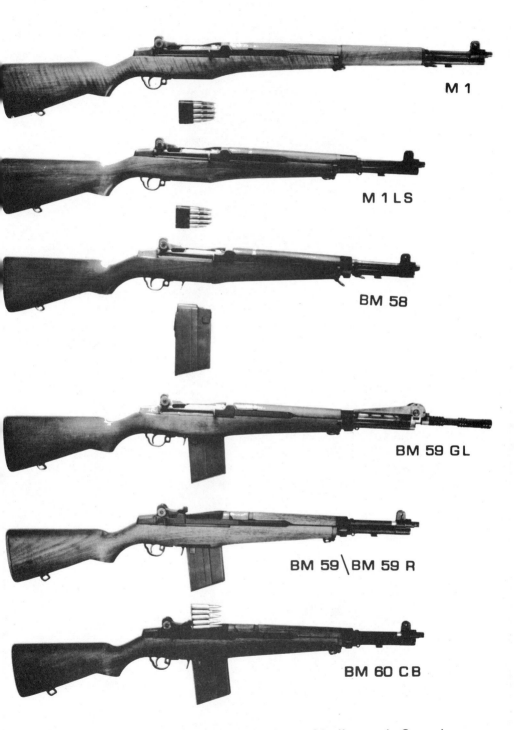

The six guns above show the range of Italian made Garands. The above guns are currently made by the P. Beretta

FUNCTIONING OF THE M1

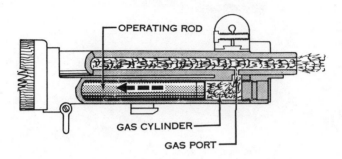

The M1 Garand is a gas operated semi automatic rifle. This means that when the cartridge is fired, some of the gas that propels the bullet is tapped off to actuate the mechanism. Basically, what happens is that when the bullet gets about 2 inches from the muzzle, it passes over a hole in the underside of the barrel. Some of the high pressure gasses rush down the hole and push against the piston end of the operating rod. The operating rod in turn cams the bolt open, and carries it to the rear, extracting the fired case as it moves to the rear. The bolt then recocks the hammer. The operating rod return spring then pulls the operating rod forward, pulling the bolt with it. The bolt in turn chambers a fresh cartridge from the clip and the cam on the operating rod rotates the bolt into locked position. Each pull of the trigger allows the cycle to repeat itself until all eight cartridges are fired. When the last cartridge is fired, the mechanism senses it and ejects the empty clip with its characteristic ping, holding the bolt back until a fresh clip of eight rounds is inserted.

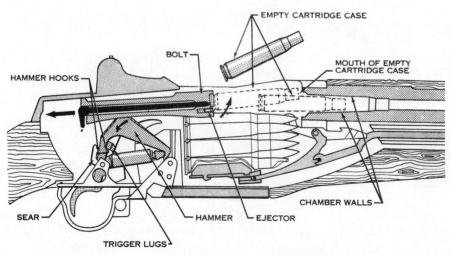

Ejecting and cocking.

FIRING MECHANISM

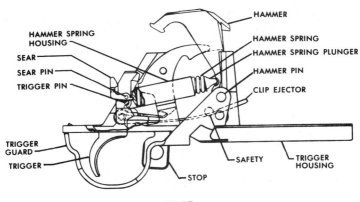

FIRED

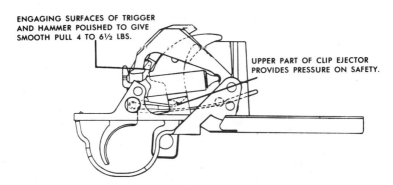

READY

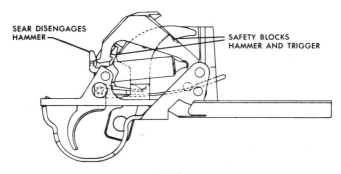

SAFE

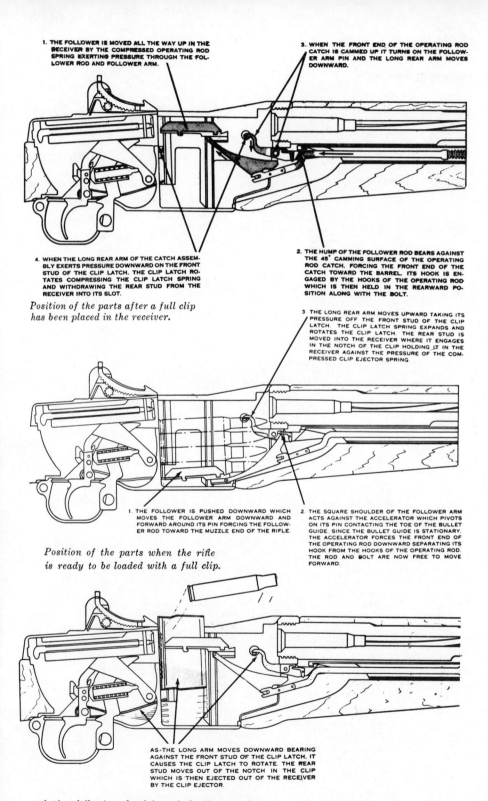

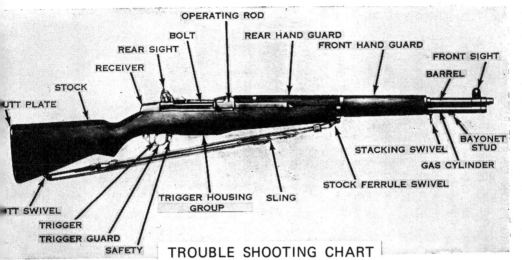

TROUBLE SHOOTING CHART

Malfunction	Cause	Correction by soldier
Failure to chamber	Dirty or rough chamber.	Clean chamber.
	Restricted gas port.	Clean gas port.
	Dirty or improperly lubricated rifle.	Clean and lubricate rifle.
	Bent clip.	Replace clip.
	Ruptured cartridge case in chamber.	Remove ruptured cartridge case.
Failure to fire (hammer releases but rifle does not fire).	Bolt not seated and locked.	Pull operating rod handle halfway to rear and release it. Insure complete locking.
	Defective or broken firing pin.	Replace firing pin.
	Defective ammunition.	Discard round.
Failure to extract	Dirty or rough chamber.	Clean chamber.
	Restricted gas port.	Clean gas port.
	Dirty ammunition.	Discard or clean round.
	Failure to replace extractor plunger and spring.	Replace extractor plunger and spring.
	Broken extractor.	Replace extractor.
Clip jumps out on seventh round.	Bent follower rod.	Replace follower rod.
Fires in bursts of two or three rounds.	Sear broken or worn, or remains in open position.	Replace trigger assembly or hammer spring housing.
	Hammer spring housing improperly assembled.	Disassemble and assemble trigger housing group correctly.
Safety releases when pressure is applied on trigger.	Worn trigger stop on safety or broken safety.	Replace safety.
Pressure on trigger does not release hammer.	Deformed hammer or trigger or worn trigger pin.	Replace defective part.
	Trigger strikes trigger housing.	Turn in to ordnance.
Creep in trigger	Burrs on trigger lugs or hammer hooks.	Replace trigger, hammer, or both.

LOADING AND UNLOADING THE M1

Loading the M1 is simple if your ammunition happens to be in Garand clips. If your cartridges are loaded correctly in an M1 eight round clip, simply pull back the operating rod until the bolt remains open and press in the entire package. It will click into place and the bolt will run forward chambering the first cartridge. When you shoot the eight rounds, the clip will be ejected. This is great if you need the full eight rounds, but say you fire five, you are left with three rounds and no way of refilling the gun unless you open the bolt ejecting the round in the chamber and pressing the clip release while trying to catch the clip and two loose cartridges.

A magazine fed gun can be replenished without emptying the chamber, and during and after WWII, attempts were made to redesign the Garand to accept a magazine.

To load the M1, pull the operating rod to the rear until it locks. Insert a clip of 8 cartridges and get your thumb clear. The bolt will run forward and chamber the first cartridge.

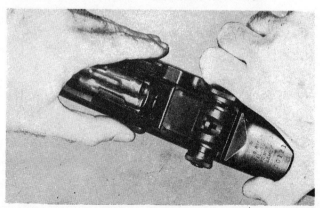

To unload the M1 pull the operating rod to the rear and hold it back. Press the clip release on the left rear of the receiver. Hold your hand over the receiver to catch the clip and cartridges. They are under heavy spring tension.

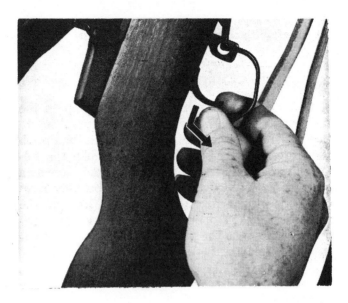

To field strip the M1 first be sure the gun is empty. Next place the butt against your thigh and pull the rear of the trigger guard, up and back, unlocking it from the trigger housing. Then lift the trigger assembly free of the stock.

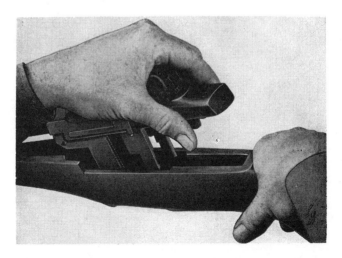

To separate the barrel and receiver from the stock, sit the rifle on a flat surface and grasp the receiver as shown. Next hit the small of the stock a light blow with your hand once or twice until the receiver rotates up and out of the stock as shown.

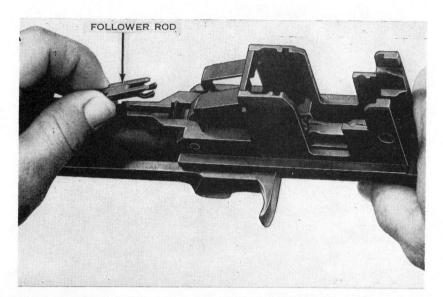

To remove the follower rod and operating rod spring, grasp the follower at the knurled position and push it back against the spring. When the fork is free pull the follower and spring out of the operating rod.

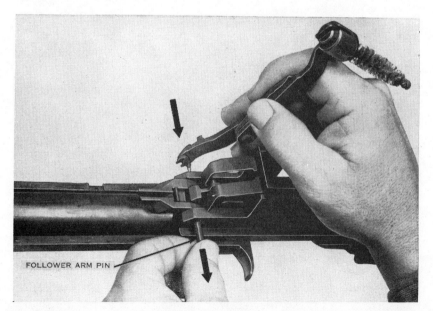

The follower arm is hinged to the receiver. To remove the arm pin use the point of a cartridge or the pin on the combination tool to push out the hinge pin as shown.

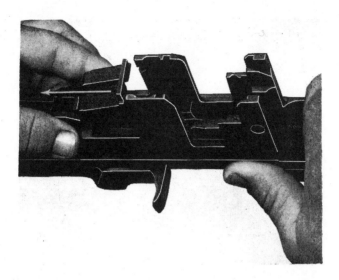

When the Follower Arm Pin is removed the Bullet Guide, Follower Arm and Operating Rod Catch can be removed as shown. Then the follower can be lifted out of the receiver.

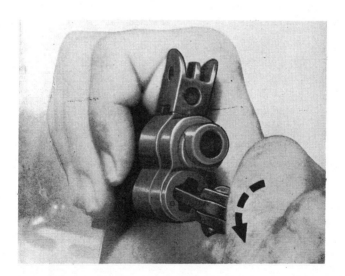

To remove the gas cylinder first unscrew the gas cylinder lock screw with a large screw driver or the combination tool supplied with the rifle. If the gas cylinder is fouled with carbon it may be necessary to tap the bayonet lug with a brass hammer or piece of wood to get it off the barrel.

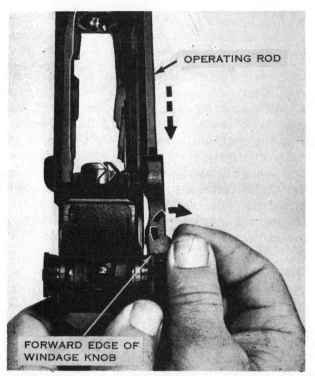

The operating rod may now be removed by pulling it back to the point shown above. Then pull the rod to the right to free it from the track in the receiver.

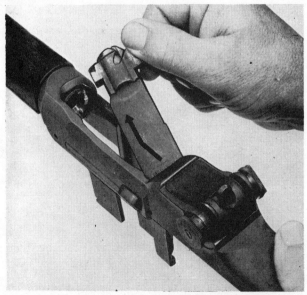

Holding the bolt by the roller on the bolt, slide it fully to the rear, then slide it forward lifting upward and outward to the right as shown. Rotate it slightly to help free it.

The M1 Garand was chambered for the 30—06 cartridge. After the second WW, a small quantity was manufactured in 30 Nato caliber for military tests. A much larger quantity were converted by the Navy (pg 46) to the 30 cal Nato cartridge. The following pages illustrate the common types of 30—06 cartridges that turn up

Two additional types of military cartridges are not illustrated. The Grenade launching cartridge with its bulletless star crimped case and the frangeable bullet M22 style. The frangeable bullet consists of 50% lead and 50% bakelite. It is mottled in color and the bullet has a green tip with a white band. It was designed to break up upon hitting 3/16" aircraft dural, and to be used in the T9 aircraft training machine gun.

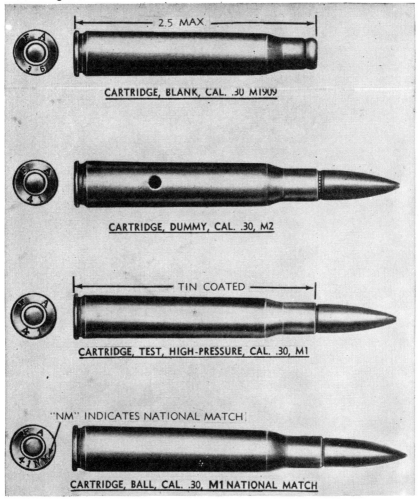

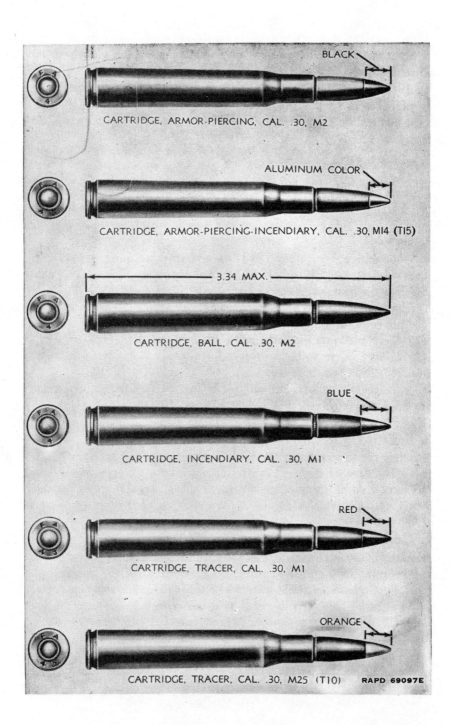